Probability

Name _____

Probability is the likelihood that a particular event or occurrence will take place. Probability is expressed as a ratio in fraction form. The probability ratio compares the number of favorable outcomes to the total possible outcomes.

Example: What is the probability of a coin landing heads up on one toss? There are **two** sides to the coin so there are **two** possible outcomes to the toss. There is **one** favorable outcome – heads! The probability is 1 out of 2 **or** 1/2.

The letters of the word "probability" are put in a bag. Find the probability of picking each letter.

1. P _____

2. R _____

3. O _____

4. B _____

5. A _____

6. I _____

7. L _____

8. T _____

9. Y _____

EXTENSION

Find the sum of the probabilities in 1 - 9 above. What do you notice? Explain.

_____

_____

Probability

Name _____

The following chart displays statistics based on a class of 26 students. The teacher has asked students to respond to a question. Using the chart, what is the probability that the following students will raise their hands?

Classroom Statistics

Groups	Out of 26 Students
Girls	14
Boys	12
Boys wearing tennis shoes	6
Girls wearing tennis shoes	9
Students wearing glasses	5
Students wearing watches	12

1. a girl _____

2. a boy _____

3. a boy wearing tennis shoes _____

4. a girl wearing tennis shoes _____

5. a student wearing glasses _____

6. a student wearing a watch _____

7. a student not wearing tennis shoes _____

EXTENSION

1. Find the sum of the probabilities in numbers 1 and 2 above. What do you notice? Explain.

   _____

   _____

2. Which has the greater probability of happening, a student wearing tennis shoes answering the question, or a student not wearing tennis shoes answering the question? Explain.

   _____

   _____

Probability

Name _____

What is the probability of rolling the following number situations with a single die? Express each answer in lowest terms.

1. even number _____

2. odd number _____

3. a number less than 6 _____

4. a number greater than 4 _____

5. a number less than 1 _____

6. a number greater than 1 _____

With a standard deck of fifty-two cards, what is the probability of choosing each of the following cards? Express each answer in lowest terms.

1. a queen _____

2. a red card _____

3. a king or a jack _____

4. a red ten _____

5. a six of diamonds _____

6. a six, seven, or eight of any suit _____

7. a nine of diamonds or hearts _____

8. a spade _____

Probability

A canister contains 200 jellybeans. There are 75 cherry-flavored, 36 lime-flavored, 44 grape-flavored, and 45 coconut-flavored jellybeans.

What is the probability of choosing a given flavor? Express each answer in lowest terms.

1. grape   _____

2. coconut   _____

3. cherry   _____

4. lime   _____

5. grape or lime   _____

6. coconut or cherry   _____

7. lemon   _____

8. grape or cherry   _____

EXTENSION

Probability ranges from 0, an impossible occurrence, to 1, an event that is certain to occur. Add together the answers to problems 1 through 4 above. What do you notice? Why is that?

_____

_____

If someone eats all of the cherry-flavored jellybeans, explain what happens to the probability of choosing each of the remaining flavors.

_____

_____

Tre... nd Co... ts

Possible outcomes can be illustrated as **trees** to determine probabilities in compound events or occurrences. Assume that the two spinners below are spun at the same time. What is the probability that each combination will be spun? Complete the **tree diagram**, then list all the possible outcomes of the two spinners and the probability of each outcome. The first two have been done for you.

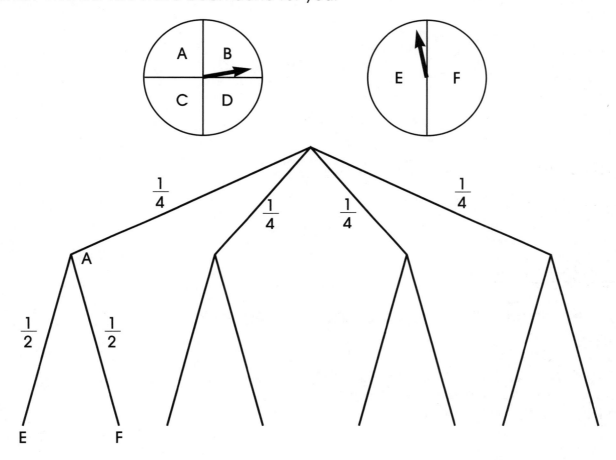

POSSIBLE COMBINATIONS: *AE, AF,* _____

PROBABILITIES: $\frac{1}{8}$, $\frac{1}{8}$, _____

EXTENSION

On another paper, draw a tree diagram using the following information. Then list all the outcomes and probabilities.

The music store offers three types of music: jazz, rock, and blues. Selections may be purchased in either cassette or CD formats.

Tree Diagrams and Compound Events

Name _____

Draw a tree diagram to illustrate each probability.

Each high school student must sign up for one foreign language course and one music course. The language choices are French, Spanish, German, or Latin. The music choices are choir, symphony, or band. List all the possible outcomes.

POSSIBLE OUTCOMES: _____

What is the probability that Latin and band will be chosen?_____

You have a coin and a spinner with the colors red, white, and blue. What is the probability of tossing heads with the color blue?

_____

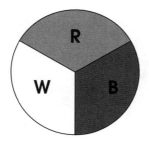

Tree Diagrams and Compound Events

Draw a tree diagram based on the spinners illustrated below. If the two spinners were each spun once, what would all the possible combinations of the two spinners be? List all the possible outcomes and the probability of each outcome.

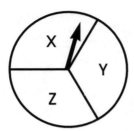 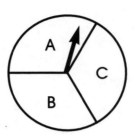

POSSIBLE COMBINATIONS: _____

PROBABILITIES: _____

Tree Diagrams and Compound Events

Mary's family is looking at new cars. They have narrowed it down to the following choices. The tree diagram below shows the possible outcomes.

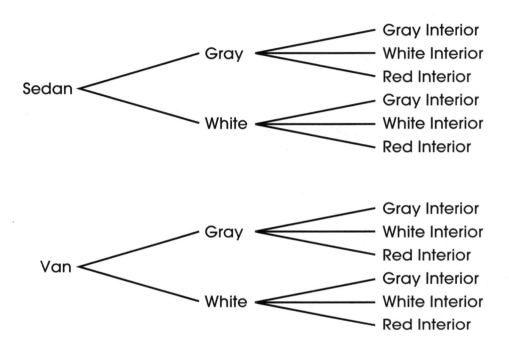

1. The compound event described above
   has how many possible outcomes?   _____

2. What is the probability that Mary's family
   will select a gray sedan with a black interior? _____

3. What is the probability that they will select
   a gray van?   _____

4. What is the probability that they will select
   a white van with a red interior?   _____

EXTENSION

On another paper, show a different way to figure the
number of possible outcomes in this compound event
without drawing a tree diagram.

Compound Events

Find the number of outcomes possible by using multiplication.

1. Roll three dice at the same time. What is the
   total number of ways the dice could land?

First Die		Second Die		Third Die		
6 choices	X		X		=	_____

2. Your combination lock has a three-digit combination.
   Each digit can be a number from one to nine.
   How many different combinations are possible?   _____

3. It is the end of the week at Camp Kukfomee. The
   cook will make sandwiches for lunch. He has four
   kinds of meats, three types of bread, two choices of
   condiments, four choices of side dishes, and three
   choices of beverages. How many different lunch
   choices can he offer?   _____

4. The car dealership in town offers 32 different models
   of vehicles. Each model has a choice of eight
   interior colors, eight exterior colors, and also the
   option of automatic or manual transmission. How
   many combinations are possible?   _____

5. The soccer team is choosing a uniform. They have a
   choice of black or white socks, black or white shoes,
   and 12 different colors of jerseys. What is the total
   number of clothing combinations?   _____

Probability and Relative Frequency

Name _____

The relative frequency of an outcome is the ratio:

$$\frac{\text{frequency of the outcome}}{\text{total frequencies of outcomes}}$$

1. Make a table to show the relative frequencies from the following car survey.

   A car dealer surveyed 200 recent car buyers. Seventy-six of his customers bought red cars, fifty-two customers bought blue cars, forty customers bought gray cars, and the rest purchased white cars.

   Color of Cars Purchased

Color	Relative Frequency

2. Using the table you made, find the probability that each car will be chosen in the future.

   a. P (red) _____   c. P (blue) _____

   b. P (grey) _____   d. P (white) _____

3. Based on the relative frequency shown on the table, what color cars should his business buy the most of next year? Explain.

   _____

4. What color cars should he buy the least of next year? Explain.

   _____

5. What else could the car dealer ask customers that would help him make the best purchasing decisions?

   _____

Probability and Relative Frequency

Name _____

Mary and Mike surveyed the 175 students at their school. Each student was asked about his/her most common lunch choice. Mary and Mike then made a frequency table.

Students at Taft Elementary	
Lunch Choice	Frequency
Pizza	92
Hamburgers	18
Spaghetti	10
Turkey Sandwiches	0
Salads	47
Bag Lunches	8

1. What is the **frequency** of students carrying bag lunches?

_____

2. What is the **relative frequency** of students carrying bag lunches?

_____

3. What is the **probability** of students choosing pizza or a bagged lunch on a given day?

_____

4. What is the **probability** of students choosing turkey sandwiches on a given day?

_____

5. Based on this survey, should turkey sandwiches still be offered? Why?

_____

6. Predict the number of students choosing pizza lunches next year in school if the school population doubles.

_____

Relative Frequency

Name _____

Solve the following.

1. Mike is an excellent offensive soccer player. On any given scoring attempt, the probability that Mike will score a soccer goal is 4/5. He has attempted to score 20 times. How many goals has he probably scored?

   _____

2. The probability that Mary will run the mile in under 5 minutes is 6/10. During the season, she has run the mile 150 times. How many times has she probably run the mile in under 5 minutes?

   _____

3. The probability that Sue's dog will catch a Frisbee when thrown is 8/12. One evening, the Frisbee was thrown 204 times. How many times did the dog probably catch the Frisbee?

   _____

4. At the movie theater, the probability that a ticket will be purchased for an action-thriller is 5/6. If 360 people go to the movie theater, how many will probably see an action-thriller?

   _____

5. The probability that a cardinal will land at the bird feeder is 2/12. Ninety-six birds have landed at the feeder today. Probably how many were cardinals?

   _____

6. The probability that a band member will play a cornet is 5/8. There are 128 band members. How many probably play a cornet?

   _____

7. The probability that a high school student will sign up for Spanish class is 2/3. There are 300 high school students. Probably how many students should the Spanish teacher count on?

   _____

8. Eight out of ten people at Ellen's school ride a bicycle for exercise. If there are 150 people, probably how many ride bicycles for exercise?

   _____

Statistics

Name _____

Using statistics is a helpful way to study various situations. The **mean** (or average) is found by dividing the sum of all possibilities by the number of possibilities. When the possibilities are arranged in numerical order, the middle one is the **median**. The possibility that occurs most frequently is the **mode**. The **range** is the difference between the greatest and the least possibility.

Mike's test scores in spelling were 94, 88, 72, 90, 70, 89, and 70.

1. What was his mean score? _____

2. What was his median score? _____

3. What was his mode score? _____

4. Which score (mean, median, mode) do you think he would like to see on his report card? Why?

   _____

5. What was the range of Mike's spelling scores? _____

The chef at Bistro Cafe found it challenging to satisfy all his diners. The ages of the diners one evening were as follows: 87, 58, 54, 61, 3, 35, 31, 28, 3, 16, and 68.

1. What is the mean age? _____

2. What is the median age? _____

3. What is the mode? _____

4. Based on the mean age, what should the chef serve, steak and lobster or macaroni and cheese? _____

5. Based on the mode, what should be served? _____

6. What is the range of the diners' ages? _____

Statistics

Find the mean, mode, median, and range of the information in each graph below.
Round the answer to the hundredths place.

Mile Relay Practice Times	
DAY	TIMES
Monday	3.29 minutes
Tuesday	3.24 minutes
Wednesday	3.48 minutes
Thursday	3.24 minutes
Friday	3.89 minutes

mean: _____

mode: _____

median: _____

range: _____

Candy Bars Sold	
DAY	NUMBER OF BARS
Monday	127
Tuesday	225
Wednesday	93
Thursday	82
Friday	111
Saturday	137
Sunday	82

mean: _____

mode: _____

median: _____

range: _____

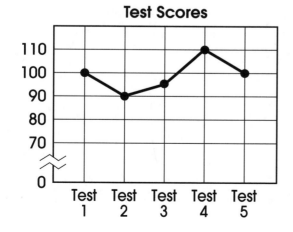

mean: _____

mode: _____

median: _____

range: _____

Statistical Experiments

Name _____

Statistical experiments involve collecting, organizing, and analyzing data.

Ms. Botanical's class is interested in growing a flower garden for the whole school to enjoy. To collect data on flower preferences, they surveyed all 435 students in the school. They noted the results below.

Favorite Flowers

Types of Flowers	Number of Votes
Black-eyed Susans	57
Lavender	63
Irises	32
Tulips	78
Hollyhocks	7
Daffodils	53
Daisies	84

Organize the Data:
List the flowers in order from the most popular to the least.

_____

Analyze the Data:
1. Based on this data, which five flowers should they plant?

_____

2. Which flower should definitely not be planted?

_____

3. Do the number of votes justify planting a garden? Why?

_____

4. What is the mean? _____

5. What is the mode? _____

6. What is the median? _____

7. What is the range? _____

Statistical Experiments

Name _____

The student council members would like to sponsor a "Fun Night" at school but are not sure what night to schedule it or whether or not the students would want one. They decided to run a survey in the school newspaper to collect data.

The results are:

Good idea 195
Bad idea 30
Monday evening 10
Tuesday evening 17   Total School Population: 237
Wednesday evening 0
Thursday evening 53
Friday evening 145

Display this information in an appropriate graph form on another paper.

Answer the following questions based on the collected data.

1. Is having a "Fun Night" a good idea? Why?

    _____

2. Which night would be the best?

    _____

3. Which night would be the second choice?

    _____

4. Do the number of votes justify the results of the survey? Explain.

    _____

5. Which night should not be chosen at all?

    _____

6. If parents were surveyed, do you think they would choose the same night? Why or why not?

    _____

Graphs

A graph compares information in a visual manner.
A **line graph** shows changes over time.

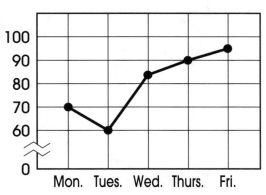

A **bar graph** shows a comparison of two or more quantities.

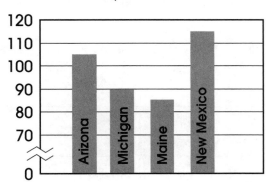

Which kind of graph would be the best way to display the following information?
Explain your choice.

1. the average monthly rainfall of Austria

_____

2. a comparison of the sales profits of five insurance companies.

_____

3. a meteor count for the month of July

_____

4. a comparison of the different depths at which 10 sea species live

_____

Graphs

Graphs have a **vertical** axis and a **horizontal** axis. The axes are labeled to show what is being compared.

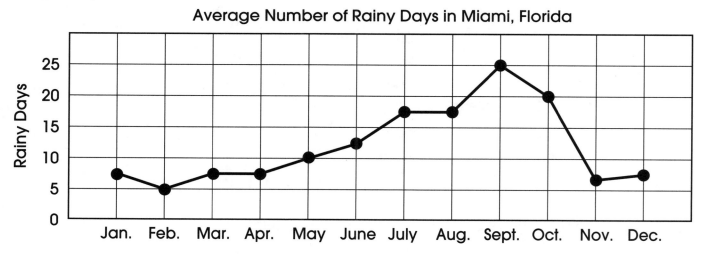

Average Number of Rainy Days in Miami, Florida

Using the data plotted on the graph, answer the following questions.

1. What is the title of the graph?

   _____

2. How is the vertical axis labeled?

   _____

3. What is contained in the horizontal axis?

   _____

4. Which month had the greatest number of rainy days?

   _____

5. Which two-month period shows the greatest change in the number of rainy days?

   _____

6. Which month was the driest?

   _____

7. Based on this graph, which two months should have been the best for tourists? Explain.

   _____

Using the graph, fill in the blanks below. (Hint: when finding the median of an even number of numerals, divide by two the sum of the two numerals in the middle.)

8. range: _____   9. mean: _____   10. median: _____   11. mode: _____

Circle Graphs

Name _____

Circle graphs are best to use when a total amount has been divided into parts. Each part illustrates a ratio of the whole.

Example:

 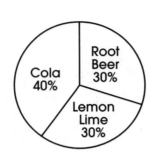

Favorite Soda Flavors

Root Beer	30%
Lemon Lime	30%
Cola	40%

Use the following information to complete the circle graphs.

1. Birthplaces of the first ten U.S. presidents:

Virginia	60%
Massachusetts	20%
New York	10%
South Carolina	10%

2. Trash collected on Ecology Day:

paper	50%
aluminum cans	15%
plastic	15%
rubber	10%
glass	10%

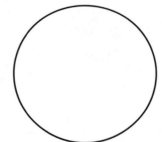

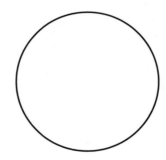

3. Pizza preferences:

cheese	30%
cheese/pepperoni	20%
cheese/mushroom	10%
deluxe	40%

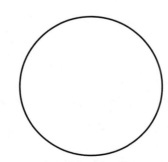

Bar Graphs

Use the following information and the boxes to create three bar graphs. Make sure to label the vertical axis, the horizontal axis, and title the graphs themselves.

1. 1993 State Populations

North Dakota	640,000
Vermont	560,000
Montana	800,000
Wyoming	450,000

2. Income per Capita

State A	$9,120
State B	$9,460
State C	$6,580
State D	$8,980

3. Heights of Garden Flowers

Daisy	3 feet 6 inches
Yarrow	2 feet
Peony	3 feet
Hollyhock	6 feet
Cone flower	3 feet

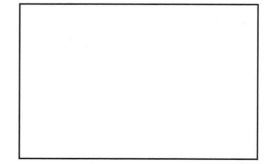

Line Graphs

Name _____

Use the following information and the boxes to create three line graphs. Make sure to label the vertical axis, the horizontal axis, and title the graphs themselves.

1. High temperatures for July 1 - 7:

Mon.	78°
Tues.	88°
Wed.	92°
Thurs.	96°
Fri.	96°
Sat.	98°
Sun.	92°

2. Cars sold in 1994:

Jan.	86
Feb.	143
Mar.	135
Apr.	152
May	201
Jun.	270
Jul.	186
Aug.	157
Sept.	164
Oct.	169
Nov.	135
Dec.	101

3. Meteor count for one week:

Mon.	17
Tues.	3
Wed.	0
Thurs.	7
Fri.	9
Sat.	8
Sun.	11

Double Line Graphs

Name _____

Double line graphing can be used to display two sets of data that will be compared over a period of time.

Example: Chris cuts lawns during the summer to earn money. Each week he cuts five lawns of different sizes, each of which takes a different amount of time. He tried to decrease his time spent on each lawn. The following chart and double line graph show his progress from week one to week four.

Lawn	Week 1	Week 4
1	1.5	1
2	3	2.5
3	1	1
4	4	3.5
5	2	.5

Week 1 ■

Week 4 ■

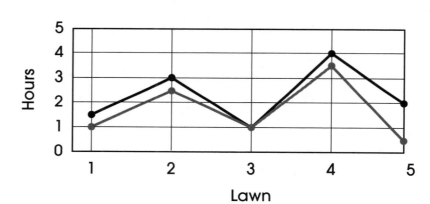

Answer the following questions using the information above.

1. Which lawn did not show a decrease in time? _____

2. Which lawn showed the greatest decrease in time? _____

3. What was the range of his time spent mowing lawns in week 4? _____

4. What was his mean time spent mowing lawns during week 1? _____

5. What was his mean time spent mowing lawns during week 4? _____

Double Line Graphs

Name _____

The volleyball players at Newhall High School were working on improving their serves. Each team member was required to practice serving 100 times each week. The table below gives the frequency of successful serves for each team member for weeks one and six. Use the frequency table to fill in the relative frequency table showing each girl's successful serves in weeks one and six.

FREQUENCY		
Name	**Week 1**	**Week 6**
Susan	38	95
Mary	72	95
Jody	40	60
Jessica	52	73
Carrie	34	72
Natasha	78	86

RELATIVE FREQUENCY		
Name	**Week 1**	**Week 6**
Susan		
Mary		
Jody		
Jessica		
Carrie		
Natasha		

Construct a double line graph showing the number of each girl's successful serves during the first and sixth weeks of practice. Remember to make a legend indicating the weeks.

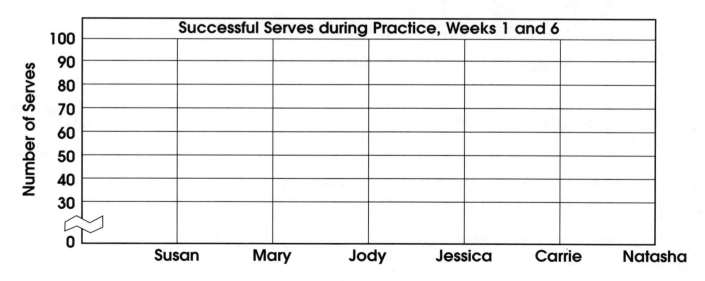

Double Bar Graphs

Name _____

Double bar graphs allow more than one set of data to be compared. The following double bar graph compares the growth between two states. (Rounded to the nearest half million)

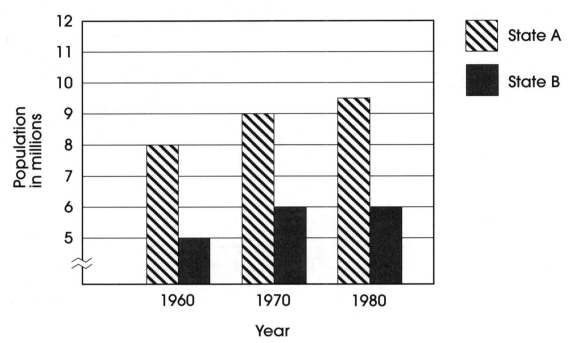

Use the double bar graph above to answer the following questions.

1. What was the population of State A in 1960?

_____

2. What was the population of State B in 1960?

_____

3. Which state experienced the greatest growth in population from 1970 to 1980?

_____

4. What was the growth of State A from 1960 to 1970?

_____

5. What was State B's population gain from 1960 to 1970?

_____

6. Which state had the greatest population growth from 1960 to 1980? What was it?

_____

Which Graph?

Tell if you would use a circle, bar, double bar, line, or double line graph to illustrate each situation.

1. A group of students are surveyed about their favorite school subjects.

    _____

2. A group of adolescents and a group of adults are surveyed about their favorite of four local radio stations.

    _____

3. The entire eighth-grade class voted on their class trip. Each person got one vote. There were five choices.

    _____

4. You wanted to keep track of your math test scores for ten weeks.

    _____

5. A travel agent wants to compare average monthly temperatures for Mexico City and Buenos Aires.

    _____

6. A meteorologist is tracking the average rainfall of Michigan for an entire year.

    _____

7. You want to compare the size of the labor force of your state with that of another state.

    _____

8. At the end of the school year, the staff at Wilson High School wanted to compare the absenteeism of the past year with the prior year on a monthly basis.

    _____

Collecting and Reporting Data

Name _____

Follow the steps to collect statistical data, design a graph, and report the results.

1. Choose one of the following topics for data collection or come up with an idea of your own. Suggestions: variety of shoes in the classroom; number of heads and tails on several coin tosses; add the roll of two dice and record a large number of trials; the number of people in your school who bought pizza for lunch over the past week.

2. Make a hypothesis. _____
   _____

3. List the raw data here. Take a large sample for accuracy.

   _____

   _____

   _____

4. Organize your data in an appropriate graph labeling all the parts accurately.

   [blank graph box]

5. Fill in these blanks.

   mean: _____   median: _____   mode: _____   range: _____

6. Write a paragraph describing the results of your data. Include a prediction for future trials.

   _____

   _____

   _____

   _____

   _____

Bar Graphs and Predictions

Name_____

City Hall is considering a proposal to build a new shopping mall in a wooded area within the city boundaries. Before it is put to a vote, City Hall decided to run a survey. Out of 4,250 residents and 182 local businesses, 1,000 people were surveyed. The results are graphed below.

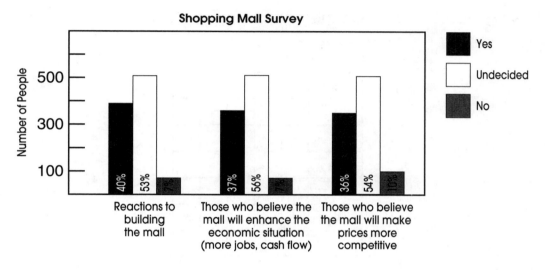

Shopping Mall Survey

Based on the information in the graph, answer the following questions.

1. Should the mall be built? Explain.

_____

_____

2. Give two reasons why City Hall should vote in favor of the shopping mall proposal.

_____

_____

3. Give two reasons why City Hall should vote against the shopping mall proposal.

_____

_____

4. What is the mean of the percentage of people who voted yes on the three questions?

_____

Coordinate Planes

Name _____

The location on a coordinate plane is given by an ordered pair. The first number (x-axis) shows the horizontal distance from the point of intersection. The second number (y-axis) shows the vertical distance from the point of intersection.

Plot the following ordered pairs. Connect the points in order.

1. (5,5)
    (4,4)
    (3,3)
    (2,2)
    (1,1)
    (0,0)

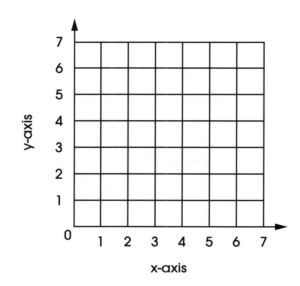

2. (1,1)
    (1,6)
    (5,6)
    (5,1)
    (1,1)

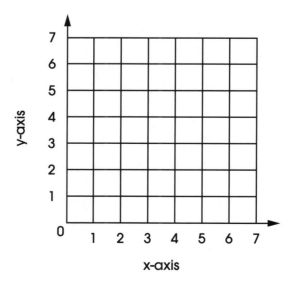

Name what is formed in number 1. _____

Name the figure formed in number 2. _____

Graphing Coordinate Planes

Name _____

Write an ordered pair for each point on the graph.

1. _____

_____

_____

_____

_____

_____

_____

_____

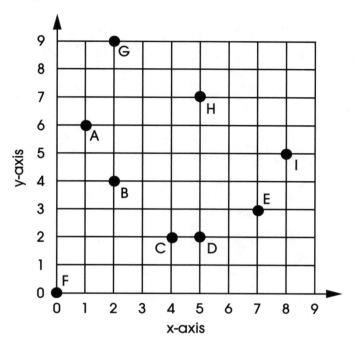

2. _____

_____

_____

_____

_____

_____

_____

_____

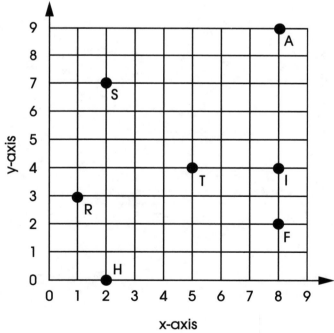

Riddle! Connect the coordinates on graph two in the following order to spell the answer to this riddle: *What is the most famous fish in Hollywood?*
(2,7) (5,4) (8,9) (1,3) (8,2) (8,4) (2,7) (2,0)

Graphing Ordered Pairs

Name _____

A. Plot each point on the graph. Connect in alphabetical order.

1. A (0, 3)
2. B (1, 5)
3. C (3, 5)
4. D (5, 5)
5. E (4, 3)
6. F (2, 3)

Name the shape. _____

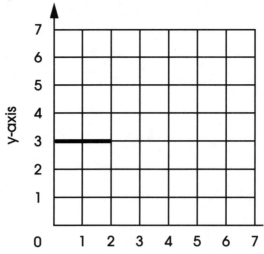

B. Plot each point on the graph.

1. m (1, 5)
2. n (2, 3)
3. o (5, 1)
4. p (1, 2)
5. q (4, 7)
6. r (3, 4)
7. s (4, 0)
8. t (0, 0)

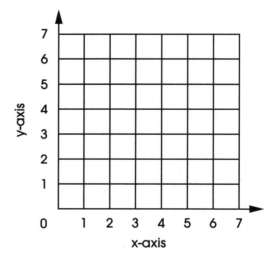

C. Which points on the graph have the following coordinates:

1. (3, 1) _____

2. (4, 5) _____

3. (2, 3) _____

4. (1, 5) _____

5. (5, 0) _____

6. (7, 6) _____

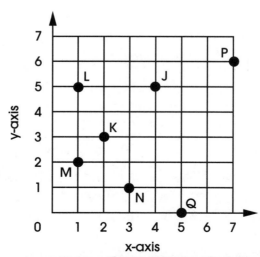

Graphing Ordered Pairs

Name _____

The x and y axes intersect at the **origin**. **Positive coordinates** are to the right and up from the origin. **Negative coordinates** are to the left and down from the origin.

Example: (-2, +1) means go left on the x-axis two units and then go up the y-axis one unit.

Point A   (-4, -1)
Point B   (-2, +1)
Point C   (+3, +1)
Point D   (+2, -1)

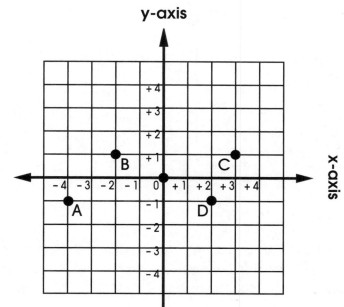

Plot the following points.

1. A   (-4, +1)
2. B   (0, -3)
3. C   (0, -4)
4. D   (-2, -4)
5. E   (+2, -3)

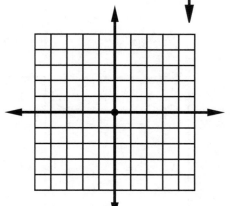

Which point has the following coordinates?

1. (-3, +4) _____
2. (0, -2) _____
3. (0, +2) _____
4. (-2, -2) _____
5. (-4, 0) _____

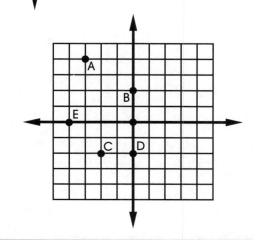

Graphing Ordered Pairs

Name _____

Draw the other half of these symmetrical figures. Then list the ordered pairs.

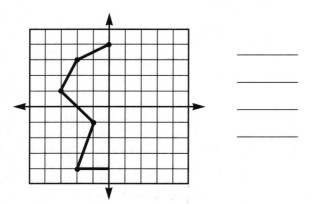

_____
_____
_____
_____
_____

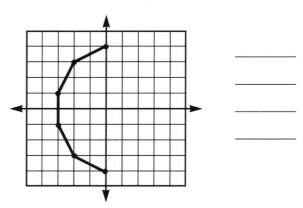

_____
_____
_____
_____
_____

Trivia Time! What planet has surface winds that have been measured at 1,500 mph — the strongest in the solar system? Find the answer to the trivia question by matching the following ordered pairs with the matching points on the graph below.

Answer: _____

(-1, 1), (-2, -4), (3, 5), (-3, -1), (0, 0), (-1, 1), (-2, -4)

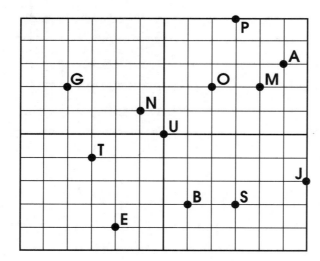

Use the same graph to answer this question.
What physical phenomena are you experiencing if you have horripilation?

_____

(-4, +2), (+2, 2), (+2, +2), (+3, -3), (-2, -4) (+1, -3), (0, 0), (+4, +2), (+3, +5), (+3, -3)

Riddle Graphing

Answer each riddle by writing the letters of the points on the graph in the same order as the ordered pairs.

Riddle: What occurs once in every minute, twice in every moment, but not once in a thousand years?

(-4, +2), (-3,-1), (-2, -4) (-1, +1), (-2, -4), (-4, +2), (-4, +2), (-2, -4), (0, 0) (1, -3)

Riddle: What gets wetter as it dries?

(+2, +2), (+3, -3), (+2, +2), (-2, -4), (0, 0) (-4, +2), (+4, +2), (+3, +5), (-2, -4), (-1, +1)

Riddle: What five-letter word has six left when you take away two?

(+5, +3), (+6, -2), (-5, -4), (-4, +2), (-6, +4)

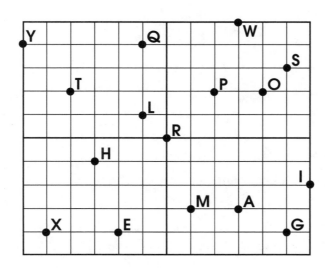

Coordinate Planes

The x and y axes separate the coordinate plane into four quadrants that are labeled I, II, III, and IV.

List the points that are located in quadrant IV.

_____

List the points that are located in quadrant II.

_____

Which points lie within quadrant I?

_____

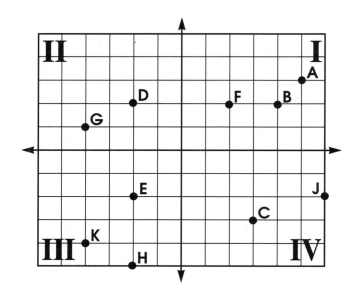

The first number of an ordered pair is called the **abscissa**. The second number of an ordered pair is called the **ordinate**.

Answer the following questions using the graph above. Name the quadrant that contains the points described.

1. The abscissa is 5. _____

2. The ordinate is 1. _____

3. The abscissa equals the ordinate. _____

4. The ordinate is -4. _____

5. The abscissa is two more than the ordinate. _____

Coordinate Planes

Name _____

Three vertexes of a rectangle are given to you. Find the fourth vertex by plotting each rectangle on the graph at the bottom of the page.

1. (-3, -3), (-7, -5), (-7, -3)   _____

2. (-1, +1), (-5, +4), (-1, +4)   _____

3. (-1, -2), (+4, -2), (+4, +1)   _____

4. (0, -3), (0, 0), (-4, -3)   _____

5. (+3, +1), (+3, +3), (+7, +1)   _____

6. (0, +4), (-5, -1), (-2, -4)   _____

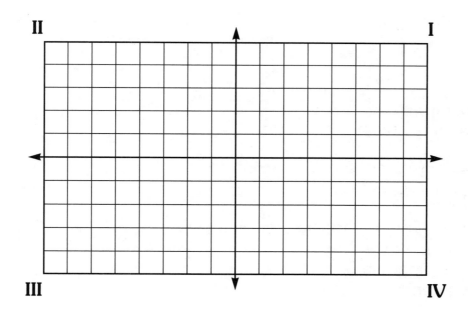

Which quadrant is rectangle number 1 in? _____

Which quadrant is rectangle number 5 in? _____

Coordinate Plane Vocabulary

Name _____

Write the vocabulary word for each definition to solve the crossword puzzle.

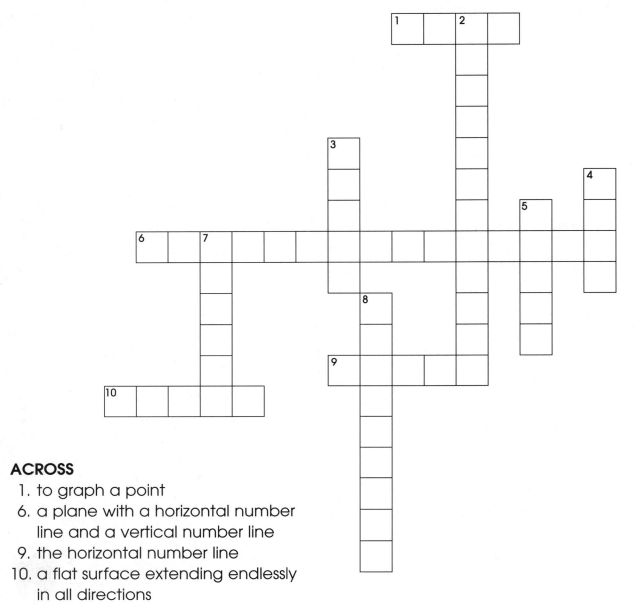

ACROSS

1. to graph a point
6. a plane with a horizontal number line and a vertical number line
9. the horizontal number line
10. a flat surface extending endlessly in all directions

DOWN

2. pairs of numbers used to locate points in a coordinate plane.
3. an exact location
4. plural of axis
5. the vertical number line
7. the point of intersection (0) of the x and y axes in a coordinate plane
8. the four sections created by the coordinate axes

Ordered Pairs in Tables

Name _____

Ordered pairs can also be displayed in a table. The ordered pairs can then be plotted on a graph as in the example.

Example:

Add 1 to x to find y.

y = x + 1	
x	y
0	1
1	2
2	3
4	5

The ordered pairs from the table:

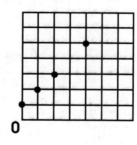

0

Follow the rule to solve for y.

y = x + 2	
x	y
0	
2	
4	
1	

y = 2x + 1	
x	y
3	
11	
4	
5	

y = 3x – 1	
x	y
8	
9	
11	
12	

y = 2x	
x	y
2	
3	
6	
10	

y = x + 1	
x	y
4	
3	
9	
12	

Find the rule.

y =	
x	y
2	1
4	2
12	6
8	4

Ordered Pairs in Tables

Name _____

Follow the rules to fill in the values for each chart. Some values are provided.

1.

y = x – 3	
x	y
	-2

2.

y = 2x – 2	
x	y
0	

3.

y = x/2	
x	y
2	

4.

y = x + 4	
x	y
	4

Can you guess what the rule (equation) is for each of the following? Write your answer in equation form below each table.

x	y
1	4
2	5
10	13

x	y
5	25
6	30
8	40

Plotting Equations

Name _____

Complete the tables and then plot the equations on the graphs.

1.

y = x – 2	
x	**y**

2.

y = x + 3	
x	**y**

3.

y = x/2 + 1	
x	**y**

What do you notice about each graph?

Linear Equations and Straight Lines

Name _____

When a table of values for x and y makes a straight line on a graph, the equation is known as a linear equation.

Example:

y = 2x – 1	
x	y
2	3
3	5
-2	-5

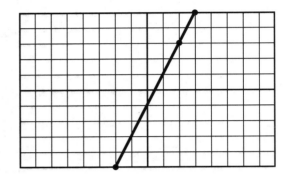

Make a table for each equation. Plot each line on the graph. State whether or not each equation is linear.

1. y = 1 - 2x _____

2. y = 2x - 3 _____

3. y = x + 0 _____

4. y = x/2 - 1 _____

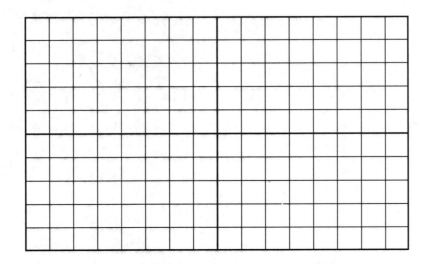

Linear Equations

Write the ordered pairs for the lines on the graph. Write the rule for each line
at the top of the table.

Line A:

y =	
x	**y**

Line B:

y =	
x	**y**

Line C:

y =	
x	**y**

Line D:

y =	
x	**y**

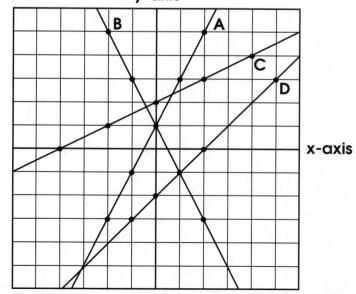

y-axis

x-axis

Answer Key

Probability, Statistics, & Graphing

Probability

Name _____

Probability is the likelihood that a particular event or occurrence will take place. Probability is expressed as a ratio in fraction form. The probability ratio compares the number of favorable outcomes to the total possible outcomes.

Example: What is the probability of a coin landing heads up on one toss? There are two sides to the coin so there are two possible outcomes to the toss. There is one favorable outcome – heads! The probability is 1 out of 2 or 1/2.

The letters of the word "probability" are put in a bag. Find the probability of picking each letter.

1. P $\frac{1}{11}$
2. R $\frac{1}{11}$
3. O $\frac{1}{11}$
4. B $\frac{2}{11}$
5. A $\frac{1}{11}$
6. I $\frac{2}{11}$
7. L $\frac{1}{11}$
8. T $\frac{2}{11}$
9. Y $\frac{1}{11}$

EXTENSION
Find the sum of the probabilities in 1 - 9 above. What do you notice? Explain.
The sum is 11 which is equal to one. All of the top numbers equal the total number.

Page 1

Probability

Name _____

The following chart displays statistics based on a class of 26 students. The teacher has asked students to respond to a question. Using the chart, what is the probability that the following students will raise their hands?

Classroom Statistics

Groups	Out of 26 Students
Girls	14
Boys	12
Boys wearing tennis shoes	6
Girls wearing tennis shoes	9
Students wearing glasses	5
Students wearing watches	12

1. a girl $\frac{14}{26}$
2. a boy $\frac{12}{26}$
3. a boy wearing tennis shoes $\frac{6}{26}$
4. a girl wearing tennis shoes $\frac{9}{26}$
5. a student wearing glasses $\frac{5}{26}$
6. a student wearing a watch $\frac{12}{26}$
7. a student not wearing tennis shoes $\frac{11}{26}$

EXTENSION
1. Find the sum of the probabilities in numbers 1 and 2 above. What do you notice? Explain.
The sum is $\frac{26}{26}$ because adding the boys and girls equals the total sampled.
2. Which has the greater probability of happening, a student wearing tennis shoes answering the question, or a student not wearing tennis shoes answering the question? Explain.
Since 15 are wearing tennis shoes and 11 are not, the probability is greater that one wearing tennis shoes will answer the question.

Page 2

Probability

Name _____

What is the probability of rolling the following number situations with a single die? Express each answer in lowest terms.

1. even number $\frac{1}{2}$
2. odd number $\frac{1}{2}$
3. a number less than 6 $\frac{5}{6}$
4. a number greater than 4 $\frac{1}{3}$
5. a number less than 1 0
6. a number greater than 1 $\frac{5}{6}$

With a standard deck of fifty-two cards, what is the probability of choosing each of the following cards? Express each answer in lowest terms.

1. a queen $\frac{1}{13}$
2. a red card $\frac{1}{2}$
3. a king or a jack $\frac{2}{13}$
4. a red ten $\frac{1}{26}$
5. a six of diamonds $\frac{1}{52}$
6. a six, seven, or eight of any suit $\frac{3}{13}$
7. a nine of diamonds or hearts $\frac{1}{26}$
8. a spade $\frac{1}{4}$

Page 3

Probability

Name _____

A canister contains 200 jellybeans. There are 75 cherry-flavored, 36 lime-flavored, 44 grape-flavored, and 45 coconut-flavored jellybeans.

What is the probability of choosing a given flavor? Express each answer in lowest terms.

1. grape $\frac{11}{50}$
2. coconut $\frac{9}{40}$
3. cherry $\frac{3}{8}$
4. lime $\frac{9}{50}$
5. grape or lime $\frac{2}{5}$
6. coconut or cherry $\frac{3}{5}$
7. lemon 0
8. grape or cherry $\frac{119}{200}$

EXTENSION
Probability ranges from 0, an impossible occurrence, to 1, an event that is certain to occur. Add together the answers to problems 1 through 4 above. What do you notice? Why is that?
They total one, an event certain to occur. Since that is all the jellybeans, you are certain to pick one of them.
If someone eats all of the cherry-flavored jellybeans, explain what happens to the probability of choosing each of the remaining flavors.
lime $\frac{36}{125}$ grape $\frac{44}{125}$ coconut $\frac{45}{125}$
The probability increases for the other flavors.

Page 4

Tree Diagrams and Compound Events

Name _____

Possible outcomes can be illustrated as trees to determine probabilities in compound events or occurrences. Assume that the two spinners below are spun at the same time. What is the probability that each combination will be spun? Complete the tree diagram, then list all the possible outcomes of the two spinners and the probability of each outcome. The first two have been done for you.

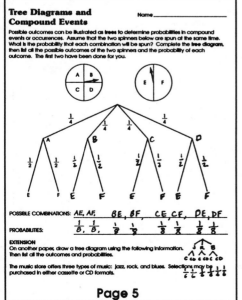

POSSIBLE COMBINATIONS: AE, AF, BE, BF, CE, CF, DE, DF

PROBABILITIES: $\frac{1}{8}, \frac{1}{8}, \frac{1}{8}, \frac{1}{8}, \frac{1}{8}, \frac{1}{8}, \frac{1}{8}, \frac{1}{8}$

EXTENSION
On another paper, draw a tree diagram using the following information. Then list all the outcomes and probabilities.
The music store offers three types of music: jazz, rock, and blues. Selections may be purchased in either cassette or CD formats.

Page 5

Tree Diagrams and Compound Events

Name _____

Draw a tree diagram to illustrate each probability.

Each high school student must sign up for one foreign language course and one music course. The language choices are French, Spanish, German, or Latin. The music choices are choir, symphony, or band. List all the possible outcomes.

POSSIBLE OUTCOMES: FC, FS, FB, SC, SS, SB, GC, GS, GB, LC, LS, LB

What is the probability that Latin and band will be chosen? $\frac{1}{12}$

You have a coin and a spinner with the colors red, white, and blue. What is the probability of tossing heads with the color blue? $\frac{1}{6}$

Page 6

Tree Diagrams and Compound Events

Name _____

Draw a tree diagram based on the spinners illustrated below. If the two spinners were each spun once, what would all the possible combinations of the two spinners be? List all the possible outcomes and the probability of each outcome.

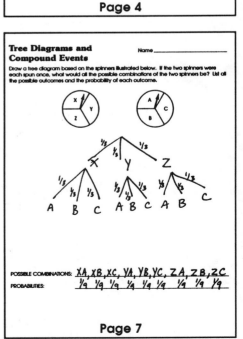

POSSIBLE COMBINATIONS: XA, XB, XC, YA, YB, YC, ZA, ZB, ZC

PROBABILITIES: $\frac{1}{9}$ $\frac{1}{9}$ $\frac{1}{9}$ $\frac{1}{9}$ $\frac{1}{9}$ $\frac{1}{9}$ $\frac{1}{9}$ $\frac{1}{9}$ $\frac{1}{9}$

Page 7

Page 8

Tree Diagrams and Compound Events

Name _____

Mary's family is looking at new cars. They have narrowed it down to the following choices. The tree diagram below shows the possible outcomes.

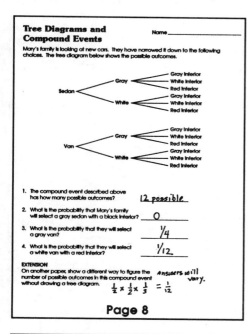

1. The compound event described above has how many possible outcomes? **12 possible**

2. What is the probability that Mary's family will select a gray sedan with a black interior? **0**

3. What is the probability that they will select a gray van? **1/4**

4. What is the probability that they will select a white van with a red interior? **1/12**

EXTENSION
On another paper, show a different way to figure the number of possible outcomes of this compound event without drawing a tree diagram.

Answers will vary.
$\frac{1}{2} \times \frac{1}{2} \times \frac{1}{3} = \frac{1}{12}$

Page 9

Compound Events

Name _____

Find the number of outcomes possible by using multiplication.

1. Roll three dice at the same time. What is the total number of ways the dice could land?

First Die	Second Die	Third Die
6 choices	x 6	x 6 = **216**

2. Your combination lock has a three-digit combination. Each digit can be a number from one to nine. How many different combinations are possible?
$9 \times 9 \times 9$ **729 combinations**

3. It is the end of the week at Camp Kuldamee. The cook will make sandwiches for lunch. He has four kinds of meats, three types of bread, two choices of condiments, four choices of side dishes, and three choices of beverages. How many different lunch choices can he offer?
$4 \times 3 \times 2 \times 4 \times 3$ **288 choices**

4. The car dealership in town offers 32 different models of vehicles. Each model has a choice of eight interior colors, eight exterior colors, and also the option of automatic or manual transmission. How many combinations are possible?
$32 \times 8 \times 8 \times 2$ **4096 combinations**

5. The soccer team is choosing a uniform. They have a choice of black or white socks, black or white shoes, and 12 different colors of jerseys. What is the total number of clothing combinations?
$2 \times 2 \times 12$ **48 combinations**

Page 10

Probability and Relative Frequency

Name _____

The relative frequency of an outcome is the ratio:

$$\frac{\text{frequency of the outcome}}{\text{total frequencies of outcomes}}$$

1. Make a table to show the relative frequencies from the following car survey.

A car dealer surveyed 200 recent car buyers. Seventy-six of his customers bought red cars, fifty-two customers bought blue cars, forty customers bought gray cars, and the rest purchased white cars.

Color of Cars Purchased

Color	Relative Frequency
red	76/200
blue	52/200
gray	40/200
white	32/200

2. Using the table you made, find the probability that each car will be chosen in the future.
   a. P (red) **19/50**   c. P (blue) **13/50**
   b. P (grey) **1/5**   d. P (white) **4/25**

3. Based on the relative frequency shown on the table, what color cars should his business buy the most of next year? Explain.
   Red, because it was purchased most frequently this year.

4. What color cars should he buy the least of next year? Explain.
   White, because the fewest customers showed interest in white cars.

5. What else should the car dealer ask customers that would help him make the best purchasing decisions?
   Answers will vary.

Page 11

Probability and Relative Frequency

Name _____

Mary and Mike surveyed the 175 students at their school. Each student was asked about his/her most common lunch choice. Mary and Mike then made a frequency table.

Students at Taft Elementary

Lunch Choice	Frequency
Pizza	92
Hamburgers	18
Spaghetti	10
Turkey Sandwiches	0
Salads	47
Bag Lunches	8

1. What is the frequency of students carrying bag lunches?
   8 people

2. What is the relative frequency of students carrying bag lunches?
   8/175

3. What is the frequency of students choosing pizza or a bagged lunch on a given day?
   100/175 → 4/7

4. What is the probability of students choosing turkey sandwiches on a given day?
   0

5. Based on this survey, should turkey sandwiches still be offered? Why?
   No, because students don't buy them.

6. Predict the number of students choosing pizza lunches next year in school if the school population doubles.
   184 students

Page 12

Relative Frequency

Name _____

Solve the following.

1. Mike is an excellent offensive soccer player. On any given scoring attempt, the probability that Mike will score a soccer goal is 4/5. He has attempted to score 20 times. How many goals has he probably scored? **16 goals**

2. The probability that Mary will run the mile in under 5 minutes is 6/10. During the season, she has run the mile 150 times. How many times has she probably run the mile in under 5 minutes? **90 times**

3. The probability that Sue's dog will catch a frisbee when thrown is 8/12. One evening, the frisbee was thrown 204 times. How many times did the dog probably catch the frisbee? **136 times**

4. At the movie theater, the probability that a ticket will be purchased for an action-thriller is 5/6. If 360 people go to the movie theater, how many will probably see an action-thriller? **300 people**

5. The probability that a cardinal will land at the bird feeder is 2/12. Ninety-six birds have landed at the feeder today. Probably how many were cardinals? **16 cardinals**

6. The probability that a band member will play a cornet is 5/8. There are 128 band members. How many probably play a cornet? **80 band members**

7. The probability that a high school student will sign up for Spanish class is 2/3. There are 300 high school students. Probably how many Spanish students should the Spanish teacher count on? **200 students**

8. Eight out of ten people at Ellen's school ride a bicycle for exercise. If there are 150 people, probably how many ride bicycles for exercise? **120 people**

Page 13

Statistics

Name _____

Using statistics is a helpful way to study various situations. The mean (or average) is found by dividing the sum of all possibilities by the number of possibilities. When the possibilities are arranged in numerical order, the middle one is the median. The possibility that occurs most frequently is the mode. The range is the difference between the greatest and the least possibility.

Mike's test scores in spelling were 94, 88, 72, 90, 70, 89, and 70.

1. What was his mean score? **82**
2. What was his median score? **88**
3. What was his mode score? **70**
4. Which score (mean, median, mode) do you think he would like to see on his report card? Why?
   The median because it is highest.
5. What was the range of Mike's spelling scores? **24**

The chef at Bistro Cafe found it challenging to satisfy all his diners. The ages of the diners that evening were as follows: 87, 58, 54, 61, 3, 35, 31, 28, 3, 16, and 68.

1. What is the mean age? **40**
2. What is the median age? **35**
3. What is the mode? **3**
4. Based on the mean age, what should the chef serve, steak and lobster or macaroni and cheese? **steak and lobster**
5. Based on the mode, what should be served? **macaroni and cheese**
6. What is the range of the diners' ages? **84 years**

Page 14

Statistics

Name _____

Find the mean, mode, median, and range of the information in each graph below. Round the answer to the hundreds place.

Mile Relay Practice Times

DAY	TIMES
Monday	3.29 minutes
Tuesday	3.24 minutes
Wednesday	3.48 minutes
Thursday	3.24 minutes
Friday	3.89 minutes

mean: **3.43**
mode: **3.24**
median: **3.29**
range: **.65**

Candy Bars Sold

DAY	NUMBER OF BARS
Monday	127
Tuesday	225
Wednesday	93
Thursday	82
Friday	111
Saturday	137
Sunday	82

mean: **122.43**
mode: **82**
median: **111**
range: **143**

Test Scores

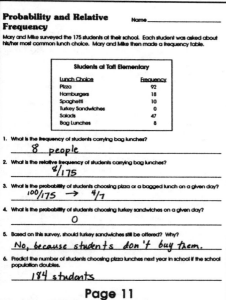

mean: **99**
mode: **100**
median: **100**
range: **20**

Page 15

Mean, Median, and Range

Name _____

Find the mean of the following series of numbers. Round the answer to the hundreds place.

1. 8, 6, 7, 10, 6 **7.4**
2. 21.6, 18.9, 31.7, 43.2 **28.85**
3. 2.7, 4.6, 4.1, 5.2 **4.15**
4. 15 kg, 17 kg, 29 kg, 21 kg **20.5 kg**
5. 67, 27, 19, 37, 42, 51 **40.5**
6. 88%, 95%, 75%, 93%, 100% **90.2%**

Find the median:

7. 21, 33, 16, 41, 19 **21**
8. 4.56, 8.19, 7.34, 7.39, 8.75 **7.39**
9. 64, 76, 62, 95, 83, 59, 63 **64**
10. 31.4 km, 12.1 km, 43 km, 34.1 km, 21 km **31.4 km**
11. $51, $36, $78, $72, $63 **$63**
12. 875, 678, 486, 364, 586, 689, 936 **678**

Find the range :

13. 2.26, 2.42, 3.31, 3.35, 2.65 **1.09**
14. 536, 990, 495, 295, 495 **695**
15. 99.8, 101.3, 96.7, 98.6 **4.6**
16. 14, 13, 7, 9, 19 **12**
17. 95%, 49%, 99%, 78%, 86% **50**
18. 17.3 m, 8.7 m, 4.5 m, 5.4 m **12.8**

Page 16

Mean, Mode, Median, and Range

Name _____

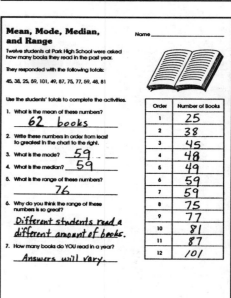

Twelve students at Park High School were asked how many books they read in the past year.

They responded with the following totals:

45, 38, 25, 59, 101, 49, 87, 75, 77, 59, 48, 81

Use the students' totals to complete the activities.

1. What is the mean of these numbers? **62 books**

2. Write these numbers in order from least to greatest in the chart to the right.

3. What is the mode? **59**

4. What is the median? **59**

5. What is the range of these numbers? **76**

6. Why do you think the range of these numbers is so great?
   Different students read a different amount of books.

7. How many books do YOU read in a year?
   Answers will vary.

Order	Number of Books
1	25
2	38
3	45
4	48
5	49
6	59
7	59
8	75
9	77
10	81
11	87
12	101

IF5120 Probability, Statistics, & Graphing

Page 17

Statistical Experiments Name _____

Statistical experiments involve collecting, organizing, and analyzing data.

Ms. Botanical's class is interested in growing a flower garden for the whole school to enjoy. To collect data on flower preferences, they surveyed all 435 students in the school. They noted the results below.

Favorite Flowers

Types of Flowers	Number of Votes
Black-eyed Susans	57
Lavender	63
Irises	32
Tulips	78
Hollyhocks	7
Daffodils	53
Daisies	84

Organize the Data:
List the flowers in order from the most popular to the least. *Daisies 84,*
Tulips 78, Lavender 63, B-E Susans 57, Daffodils 53,
Irises 32, Hollyhocks 7

Analyze the Data:
1. Based on this data, which five flowers should they plant?
Daisies, tulips, lavender, black-eyed susans, daffodils
2. Which flower should definitely not be planted?
hollyhocks
3. Do the number of votes justify planting a garden? Why?
Not really, the survey wasn't asking that question.
4. What is the mean? *53.43*
5. What is the mode? *No mode*
6. What is the median? *57*
7. What is the range? *77*

Page 17

Page 18

Statistical Experiments Name _____

The student council members would like to sponsor a "Fun Night" at school but are not sure what night to schedule it on or whether or not the students would want one. They decided to run a survey in the school newspaper to collect data.

The results are:

Good idea 195
Bad idea 30
Monday evening 10
Tuesday evening 17   Total School Population: 237
Wednesday evening 0
Thursday evening 53
Friday evening 145

Display this information in an appropriate graph form on another paper. *Graphs will vary.*

Answer the following questions based on the collected data.
1. Is having a "Fun Night" a good idea? Why?
Yes, more than half the students think it is a good idea.
2. Which night would be the best?
Friday evening had the most votes.
3. Which night would be the second choice?
Thursday evening
4. Do the number of votes justify the results of the survey? Explain.
Yes, the majority of students voted.
5. Which night should not be chosen at all?
Wednesday evening
6. If parents were surveyed, do you think they would choose the same night? Why or why not?
Answers will vary.

Page 18

Page 19

Graphs Name _____

A graph compares information in a visual manner.
A line graph shows changes over time.

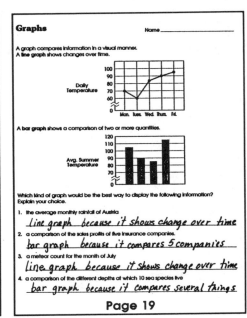

A bar graph shows a comparison of two or more quantities.

Which kind of graph would be the best way to display the following information? Explain your choice.

1. the average monthly rainfall of Austria
line graph because it shows change over time
2. a comparison of the sales profits of five insurance companies.
bar graph because it compares 5 companies
3. a meteor count for the month of July
line graph because it shows change over time
4. a comparison of the different depths at which 10 sea species live
bar graph because it compares several things

Page 19

Page 20

Graphs Name _____

Graphs have a vertical axis and a horizontal axis. The axes are labeled to show what is being compared.

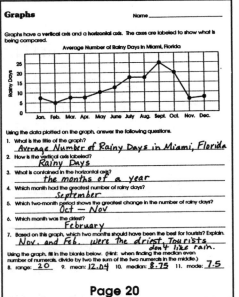

Using the data plotted on the graph, answer the following questions.
1. What is the title of the graph?
Average Number of Rainy Days in Miami, Florida
2. How is the vertical axis labeled?
Rainy Days
3. What is contained in the horizontal axis?
the months of a year
4. Which month had the greatest number of rainy days?
September
5. Which two-month period shows the greatest change in the number of rainy days?
Oct — Nov
6. Which month was the driest?
February
7. Based on this graph, which two months should have been the best for tourists? Explain.
Nov. and Feb. were the driest. Tourists don't like rain.

Using the graph, fill in the blanks below. (Hint: when finding the median even number of numerals, divide by two the sum of the two numerals in the middle.)
8. range: *20* 9. mean: *12.04* 10. median: *8.75* 11. mode: *7.5*

Page 20

Page 21

Circle Graphs Name _____

Circle graphs are best to use when a total amount has been divided into parts. Each part illustrates a ratio of the whole.

Example:

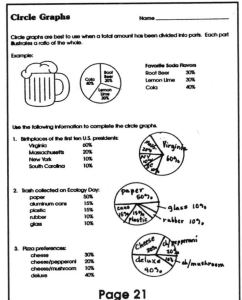

Favorite Soda Flavors
Root Beer 30%
Lemon Lime 30%
Cola 40%

Use the following information to complete the circle graphs.

1. Birthplaces of the first ten U.S. presidents:
Virginia 60%
Massachusetts 20%
New York 10%
South Carolina 10%

2. Trash collected on Ecology Day:
paper 50%
aluminum cans 15%
plastic 15%
rubber 10%
glass 10%

3. Pizza preferences:
cheese 30%
cheese/pepperoni 20%
cheese/mushroom 10%
deluxe 40%

Page 21

Page 22

Bar Graphs Name _____

Use the following information and the box to create a bar graph. Make sure to label the vertical axis, the horizontal axis, and title the graph itself.

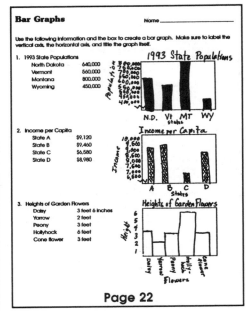

1. 1993 State Populations
North Dakota 640,000
Vermont 560,000
Montana 800,000
Wyoming 450,000

2. Income per Capita
State A $9,120
State B $9,460
State C $6,580
State D $8,980

3. Heights of Garden Flowers
Daisy 3 feet 6 inches
Yarrow 2 feet
Peony 3 feet
Hollyhock 6 feet
Cone flower 3 feet

Page 22

Page 23

Line Graphs Name _____

Use the following information and the boxes to create three line graphs. Make sure to label the vertical axis, the horizontal axis, and title the graphs themselves.

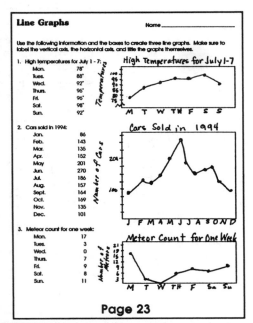

1. High temperatures for July 1-7:
Mon. 78°
Tues. 88°
Wed. 92°
Thurs. 96°
Fri. 96°
Sat. 98°
Sun. 92°

2. Cars sold in 1994:
Jan. 86
Feb. 143
Mar. 135
Apr. 152
May 201
Jun. 270
Jul. 186
Aug. 157
Sept. 169
Oct. 169
Nov. 135
Dec. 101

3. Meteor count for one week:
Mon. 17
Tues. 3
Wed. 0
Thurs. 7
Fri. 9
Sat. 8
Sun. 11

Page 23

Page 24

Double Line Graphs Name _____

Double line graphing can be used to display two sets of data that will be compared over a period of time.

Example: Chris cuts lawns during the summer to earn money. Each week he cuts five lawns of different sizes, each of which takes a different amount of time. He tried to decrease his time spent on each lawn. The following chart and double line graph show his progress from week one to week four.

Lawn	Week 1	Week 4
1	1.5	1
2	3	2.5
3	1	1
4	4	3.5
5	2	.5

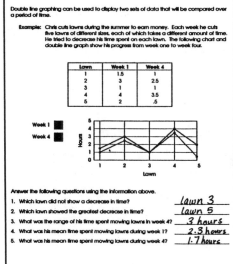

Week 1 ■
Week 4 ■

Answer the following questions using the information above.
1. Which lawn did not show a decrease in time? *lawn 3*
2. Which lawn showed the greatest decrease in time? *lawn 5*
3. What was the range of his time spent mowing lawns in week 4? *3 hours*
4. What was his mean time spent mowing lawns during week 1? *2.3 hours*
5. What was his mean time spent mowing lawns during week 4? *1.7 hours*

Page 24

Page 25

Double Line Graphs Name _____

The volleyball players at Newhall High School were working on improving their serves. Each team member was required to practice serving 100 times each week. The table below gives the frequency of successful serves for each team member for weeks one and six. Use the frequency table to fill in the relative frequency table showing each girl's successful serves in weeks one and six.

FREQUENCY		
Name	Week 1	Week 6
Susan	38	95
Mary	72	95
Jody	40	60
Jessica	52	73
Carrie	34	72
Natasha	78	86

RELATIVE FREQUENCY		
Name	Week 1	Week 6
Susan	38/100	95/100
Mary	72/100	95/100
Jody	40/100	60/100
Jessica	52/100	73/100
Carrie	34/100	72/100
Natasha	78/100	86/100

Construct a double line graph showing the number of each girl's successful serves during the first and sixth weeks of practice. Remember to make a legend indicating the weeks.

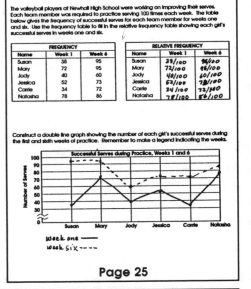

week one ———
week six - - - -

Page 25

Double Bar Graphs

Name _____

Double bar graphs allow more than one set of data to be compared. The following double bar graph compares the growth between two states. (Rounded to the nearest half million)

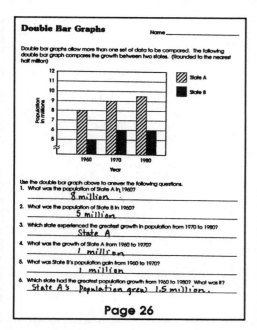

Use the double bar graph above to answer the following questions.

1. What was the population of State A in 1960?
   8 million

2. What was the population of State B in 1960?
   5 million

3. Which state experienced the greatest growth in population from 1970 to 1980?
   State A

4. What was the growth of State A from 1960 to 1970?
   1 million

5. What was State B's population gain from 1960 to 1970?
   1 million

6. Which state had the greatest population growth from 1960 to 1980? What was it?
   State A's population grew 1.5 million.

Page 26

Which Graph?

Name _____

Tell if you would use a circle, bar, double bar, line, or double line graph to illustrate each situation.

1. A group of students are surveyed about their favorite school subjects.
   Circle graph

2. A group of adolescents and a group of adults are surveyed about their favorite of four local radio stations.
   Double bar graph

3. The entire eighth-grade class voted on their class trip. Each person got one vote. There were five choices.
   circle graph

4. You wanted to keep track of your math test scores for ten weeks.
   line graph

5. A travel agent wants to compare average monthly temperatures for Mexico City and Buenos Aires.
   double line graph

6. A meteorologist is tracking the average rainfall of Michigan for an entire year.
   line graph

7. You want to compare the size of the labor force of your state with that of another state.
   bar graph

8. At the end of the school year, the staff at Wilson High School wanted to compare the absenteeism of the past year with the prior year on a monthly basis.
   double line graph

Note: Answers may vary. These are the best choices based on the previous activities (p. 19-26).

Page 27

Collecting and Reporting Data

Name _____

Follow the steps to collect statistical data, design a graph, and report the results.   Answers will vary.

1. Choose one of the following topics for data collection or come up with an idea of your own: variety of shoes in the classroom, number of heads and tails on several coin tosses, add the roll of two dice and record a large number of trials, the number of people in your school who bought pizza for lunch over the past week.

2. Make a hypothesis. _____

3. List the raw data here. Take a large sample for accuracy.
   _____
   _____
   _____

4. Organize your data in a appropriate graph labeling all the parts accurately.

5. Fill in these blanks.
   mean: ___ median: ___ mode: ___ range: ___

6. Write a paragraph describing the results of your data. Include a prediction for future trials.
   _____

Page 28

Bar Graphs and Predictions

Name _____

City Hall is considering a proposal to build a new shopping mall in a wooded area within the city boundaries. Before it is put to a vote, City Hall decided to run a survey. Out of 4,250 residents and 182 local businesses, 1,000 people were surveyed. The results are graphed below.

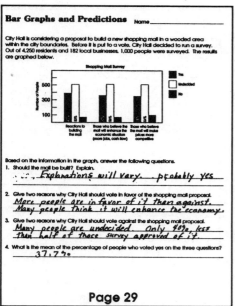

Based on the information in the graph, answer the following questions.

1. Should the mall be built? Explain.
   Explanations will vary. probably yes

2. Give two reasons why City Hall should vote in favor of the shopping mall proposal.
   More people are in favor of it than against. Many people think it will enhance the economy.

3. Give two reasons why City Hall should vote against the shopping mall proposal.
   Many people are undecided. Only 40% less than half of those survey approved of it.

4. What is the mean of the percentage of people who voted yes on the three questions?
   37.7%

Page 29

Coordinate Planes

Name _____

The location on a coordinate plane is given by an ordered pair. The first number (x-axis) shows the horizontal distance from the point of intersection. The second number (y-axis) shows the vertical distance from the point of intersection.

Plot the following ordered pairs. Connect the points in order.

1. (5,5)
   (4,4)
   (3,3)
   (2,2)
   (1,1)
   (0,0)

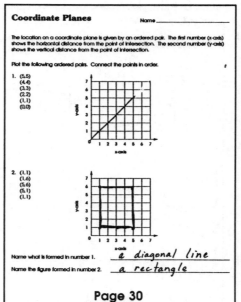

2. (1,1)
   (1,6)
   (5,6)
   (5,1)
   (1,1)

Name what is formed in number 1. a diagonal line

Name the figure formed in number 2. a rectangle

Page 30

Graphing Coordinate Planes

Name _____

Write an ordered pair for each point on the graph.

1. A (1,6)
   B (2,4)
   C (4,2)
   D (5,2)
   E (7,3)
   F (0,0)
   G (2,9)
   H (5,7)
   I (8,5)

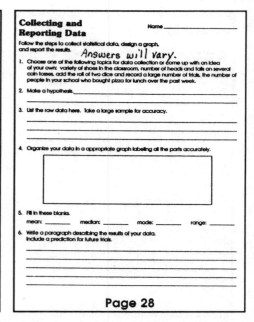

2. R (1,3)
   H (2,0)
   S (2,7)
   T (5,4)
   F (8,2)
   I (8,4)
   A (8,9)

Riddle! Connect the coordinates on graph two in the following order to spell the answer to this riddle: What is the most famous fish in Hollywood?
(2,7) (5,4) (8,9) (1,3) (8,2) (8,4) (2,7) (2,0)
   Star fish

Page 31

Graphing Ordered Pairs

Name _____

A. Plot each point on the graph below. Connect in alphabetical order.

1. A (0, 3)   4. D (5, 5)
2. B (1, 5)   5. E (4, 3)
3. C (3, 5)   6. F (2, 3)

Name the shape. parallelogram

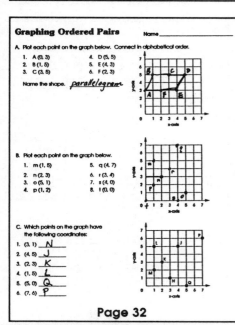

B. Plot each point on the graph below.

1. m (1, 5)   5. q (4, 7)
2. n (2, 3)   6. r (3, 4)
3. o (5, 1)   7. s (4, 0)
4. p (1, 2)   8. t (0, 0)

C. Which points on the graph have the following coordinates:

1. (3, 1) __N__
2. (4, 5) __J__
3. (2, 3) __K__
4. (1, 5) __L__
5. (5, 0) __Q__
6. (7, 6) __P__

Page 32

Graphing Ordered Pairs

Name _____

The x and y axes intersect at the origin. Positive coordinates are to the right and up from the origin. Negative coordinates are to the left and down from the origin.

Example: (-2, +1) means go left on the x-axis two units and then go up the y-axis one unit.

Point A (-4, -1)
Point B (-2, +1)
Point C (+3, +1)
Point D (+2, -1)

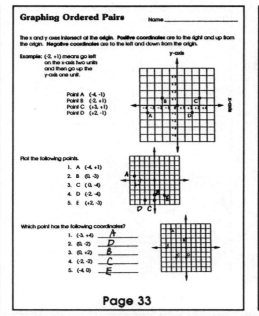

Plot the following points.

1. A (-4, +1)
2. B (1, -3)
3. C (0, -4)
4. D (-2, -1)
5. E (+2, -3)

Which point has the following coordinates?

1. (-3, +4) __A__
2. (1, -2) __D__
3. (0, +2) __B__
4. (-2, -2) __C__
5. (-4, 0) __E__

Page 33

Graphing Ordered Pairs

Name _____

Draw the other half of these symmetrical figures. Then list the ordered pairs.

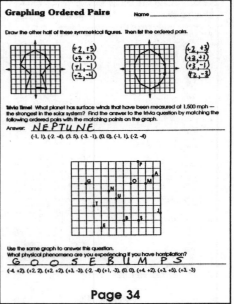

(+2, +3)
(+3, +1)
(+1, -1)
(+2, -4)

(+2, +3)
(+2, +1)
(+2, -1)
(+2, -3)

Trivia Time! What planet has surface winds that have been measured at 1,500 mph — the strongest in the solar system? Find the answer to the trivia question by matching the following ordered pairs with the matching points on the graph.

Answer: NEPTUNE
(-1, 1), (-2, -4), (3, 5), (-3, -1), (0, 0), (-1, 1), (-2, -4)

Use the same graph to answer this question.
What physical phenomena are you experiencing if you have horripilation?
G O O S E B U M P S
(-4, +2), (+2, 2), (+2, +2), (+3, -3), (-2, -4) (+1, -3), (0, 0), (+4, +2), (+3, +5), (+3, -3)

Page 34

© Instructional Fair, Inc.   47   IF5120 Probability, Statistics, & Graphing

Page 35

Riddle Graphing   Name _____

Answer each riddle by writing the letters of the points on the graph in the same order as the ordered pairs.

Riddle: What occurs once in every minute, twice in every moment, but not once in a thousand years?
(-4, +2), (-3, -1), (-2, -4) (-1, +1), (-2, -4), (-4, +2), (+4, +2), (-2, -4), (0, 0) (1, -3)
T H E   L E T T E R   M

Riddle: What gets wetter as it dries?
(+2, +2), (+3, -3), (+2, +2), (-2, -4), (0, 0) (-4, +2), (+4, +2), (+3, +5), (-2, -4), (-1, +1)
P A P E R T O W E L

Riddle: What five-letter word has six left when you take away two?
(+5, +3), (+6, -2), (-5, -4), (-4, +2), (-4, +4)
S I X T Y

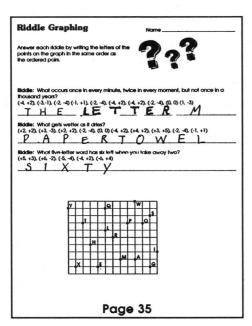

Page 35

Page 36

Coordinate Planes   Name _____

The x and y axes separate the coordinate plane into four quadrants that are labeled I, II, III, and IV.

List the points that are located in quadrant IV.
C, J

List the points that are located in quadrant II.
D, G

Which points lie within quadrant I?
A, F, B

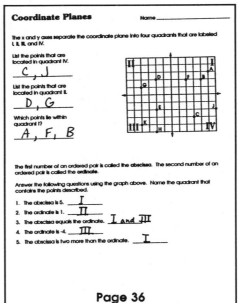

The first number of an ordered pair is called the abscissa. The second number of an ordered pair is called the ordinate.

Answer the following questions using the graph above. Name the quadrant that contains the points described.

1. The abscissa is 5.   I
2. The ordinate is 1.   II
3. The abscissa equals the ordinate.   I and III
4. The ordinate is -4.   III
5. The abscissa is two more than the ordinate.   I

Page 36

Page 37

Coordinate Planes   Name _____

Three vertexes of a rectangle are given to you. Find the fourth vertex by plotting each rectangle on the graph at the bottom of the page.

1. (-3, -3), (-7, -5), (-7, -3)   (-3, -5)
2. (-1, +1), (-5, +4), (-1, +4)   (-5, +1)
3. (-1, -2), (+4, -2), (+4, +1)   (-1, +1)
4. (0, -3), (0, 0), (-4, -3)   (-4, 0)
5. (+3, +1), (+3, +3), (+7, +1)   (+7, +3)
6. (0, +4), (-5, -1), (-2, -4)   (+3, +1)

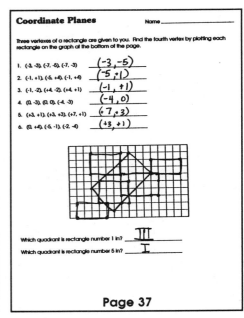

Which quadrant is rectangle number 1 in?   III
Which quadrant is rectangle number 5 in?   I

Page 37

Page 38

Coordinate Plane Vocabulary   Name _____

Write the vocabulary word for each definition to solve the crossword puzzle.

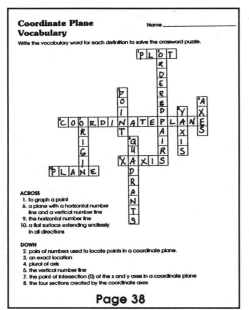

ACROSS
1. to graph a point
6. a plane with a horizontal number line and a vertical number line
9. the horizontal number line
10. a flat surface extending endlessly in all directions

DOWN
2. pairs of numbers used to locate points in a coordinate plane.
3. an exact location
4. plural of axis
5. the vertical number line
7. the point of intersection (0) of the x and y axes in a coordinate plane
8. the four sections created by the coordinate axes

Page 38

Page 39

Ordered Pairs in Tables   Name _____

Ordered pairs can also be displayed in a table. The ordered pairs can then be plotted on a graph as in the example.

Example:
Add 1 to x to find y.

y = x + 1	
x	y
0	1
1	2
2	3
4	5

The ordered pairs from the table:

Follow the rule to solve for y.

y = x + 2	
x	y
0	2
2	4
4	6
1	3

y = 2x + 1	
x	y
3	7
11	23
4	9
5	11

y = 3x - 1	
x	y
8	23
9	26
11	32
12	35

Find the rule.

y = 2x	
x	y
2	4
3	6
6	12
10	20

y = x + 1	
x	y
4	5
3	4
9	10
12	13

y = ½x	
x	y
2	1
4	2
12	6
8	4

Page 39

Page 40

Ordered Pairs in Tables   Name _____

Follow the rules to fill in the values for each chart. Some values are provided.

1.
y = x - 3	
x	y
1	-2
3	0
6	3
2	-1

2.
y = 2x - 2	
x	y
0	-2
2	2
6	10
3	4

Answers will vary.

3.
y = x/2	
x	y
2	1
8	4
4	2
6	3

4.
y = x + 4	
x	y
0	4
2	6
4	8
6	10

Can you guess what the rule (equation) is for each of the following? Write your answer in equation form below each table.

x	y
1	4
2	5
10	13

y = x + 3

x	y
5	25
6	30
8	40

y = 5x

Page 40

Page 41

Plotting Equations   Name _____

Complete the tables and then plot the equations on the graphs.

1.
y = x - 2	
x	y
2	0
3	1
4	2
-2	-4

Chart numbers will vary

2.
y = x + 3	
x	y
3	6
0	3
-3	0
2	5

3.
y = x/2 + 1	
x	y
2	2
4	3
0	1
-2	0

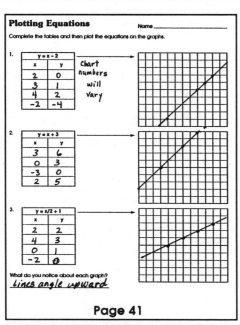

What do you notice about each graph?
Lines angle upward

Page 41

Page 42

Linear Equations and Straight Lines   Name _____

When a table of values for x and y makes a straight line on a graph, the equation is known as a linear equation.

Example:

y = 2x - 1	
x	y
1	1
2	3
3	5
-2	-5

Make a table for each equation. Plot each line on the graph. State whether or not each equation is linear.

1. y = 1 - 2x   linear
2. y = 2x - 3   linear

tables will vary

3. y = x + 0   linear
4. y = x/2 - 1   linear

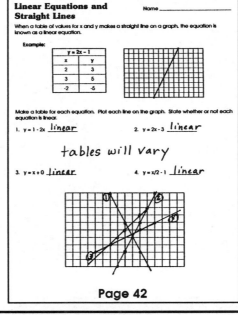

Page 42

Page 43

Linear Equations   Name _____

Write the ordered pairs for the lines on the graph. Write the rule for each line above the table.

Line A:
y = 2x + 1	
x	y
2	5
1	3
-1	-1
-2	-3
-3	-5

Line B:
y = -2x + 1	
x	y
-2	5
-1	3
0	1
1	-1
2	-3

Line C:
y = x/2 + 2	
x	y
4	4
2	3
0	2
-2	1
-4	0

Line D:
y = x - 2	
x	y
5	3
2	0
0	-1
0	-2
-1	-3

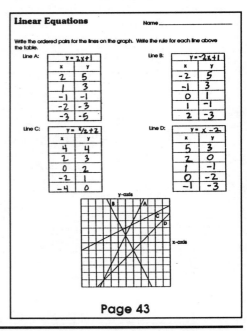

Page 43

         IF5120 Probability, Statistics, & Graphing